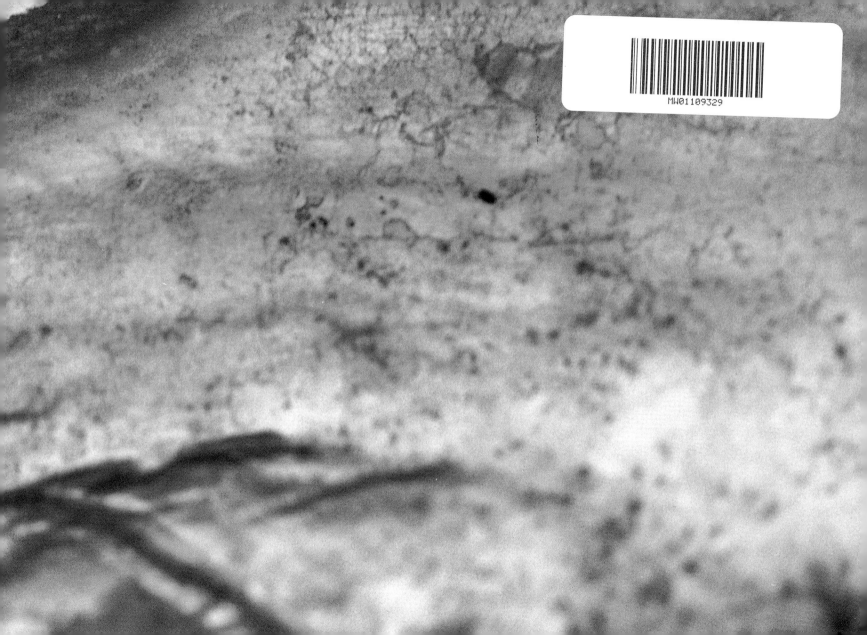

The Majesty And Mystery Of The Sea

Published by:
THIRD STORY BOOKS, 955 Connecticut Avenue, Suite 1302, Bridgeport, CT 06607

Distributed to the trade by:
ANDREWS AND MCMEEL, 4900 Main Street Kansas City, MO 64112

ISBN # 1-884506-14-3

Library of Congress Catalog Card Number 94-60732
Endpapers: Abalone Shell (Kevin and Cat Sweeney).
Design by TC&L Inc.

© 1994 Sea World, Inc. All rights reserved.
Sea World, the Whale Logo, and the Sea World Characters are trademarks of Sea World, Inc. No part of this book may be reproduced or copied in any form or by any means — graphics, electronic, or mechanical, including photo-copying, taping, or information storage and retrieval systems — without written permission of the publisher.

The editor and publisher are grateful for permission to include copyright material in this book. A listing of copyright notices and text sources can be found on page 94 — The Authors. All photographs are used courtesy of Sea World, except as noted on page 95 — The Photographers.

Although every effort has been made to secure permissions prior to printing, this has not always been possible. The publisher apologizes for any errors or omissions that may have inadvertently occurred, but if contacted will rectify these at the earliest opportunity. The publishers wish to express their thanks to Robert Rattner for his unparalleled editorial assistance.

Printed in Singapore

1 2 3 4 5 6 7 8 9 10

First Edition

The Majesty And Mystery Of The Sea

A Photographic Celebration Of The Marine Environment

Postlude By *Faith Popcorn*

EDITED BY JOHN SAMMIS

Green Sea Turtle

Prelude

The world below the brine,
Forests at the bottom of the sea, the
    branches and leaves,
Sea-lettuce, vast lichens, strange flowers and
    seeds, the thick tangle, openings, and
    pink turf,
Different colours, pale grey and green,
    purpose white, and gold, the play of light
    through water,
Dumb swimmers there among the rocks,
    coral, gluten, grass, rushes, and the
    aliment of the swimmers,
Sluggish existences grazing there
    suspended, or slowly crawling close to
    the bottom,
The sperm-whale at the surface blowing air
    and spray, or disporting with his flukes,
The leaden-eyed shark, the walrus, the
    turtle, the hairy sea-leopard, and the
    sting-ray.
Passions there, wars, pursuits, tribes, sight in
    those ocean-depths, breathing that thick-
    breathing air, as so many do,
The change thence to the sight here, and to
    the subtle air breathed by beings like us
    who walk this sphere,
The change onward from ours to that of
    beings who walk other spheres.

                      Walt Whitman

All things are water.
                    Plutarch

LIGHTHOUSE POINT, SANTA CRUZ, CA

Light is the first of painters.
Ralph Waldo Emerson

English Channel

Any naturalist who is lucky enough to travel, at certain moments has experienced a feeling of overwhelming exultation at the beauty and complexity of life and a feeling of depression that there is so much to see, to observe, to learn, that one lifetime is an unfairly short span to be allotted for such a paradise of enigmas as the world is. . .

But there is one experience, perhaps above all others, that a naturalist should try to have before he dies, and that is the astonishing and humbling experience of exploring a tropical reef. It seems that in this one action you use nearly every one of your senses, and one feels that one could uncover hidden senses as well.

Gerald Durrell

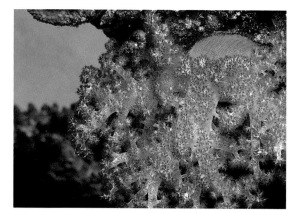

SOFT CORAL

CHRISTMAS TREE WORMS ON BRAIN CORAL

Sea Fan Coral

WEST PACIFIC CORAL REEF

Tropical reefs seethe with a multitude of diverse life. Like ever-churning engines, corals, fish, plants, and animals work non-stop day and night on the reefs. So much activity goes on that it is difficult for a visitor to know where to focus.

Ann Scarborough-Bull, PhD

MAGNIFICENT BANDED FAN WORMS ON CORAL

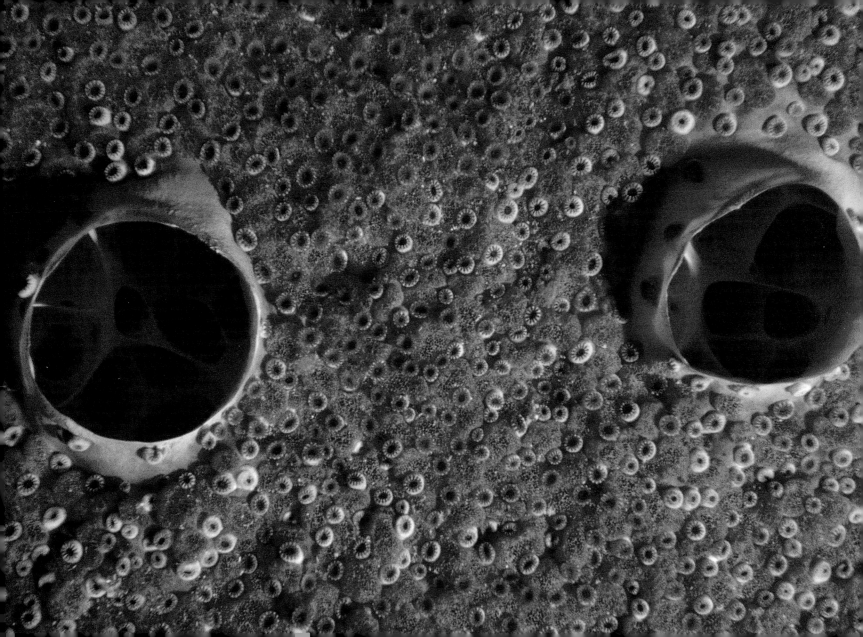

The old-fashioned boardinghouse was a grand American tradition. In its heyday, the typical establishment housed a melting pot of tenants, well-fed by the hostess at her own table. ...Sponges qualify as perhaps the most interesting and prolific boardinghouses on the reef.

— Nancy Sefton

Urchin, Corallina and Sponges

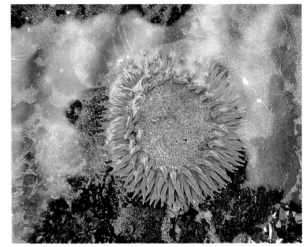

GIANT GREEN SEA ANEMONE

Once, after spending an entire dive hunched over a particularly special anemone, I turned to see my safety diver visibly shaking from the cold. When we surfaced, he said that I had spent three hours and fifteen minutes with that one anemone. "Why?" he asked. I could only answer, "Because I was trying to see it." To my eye it offered many faces, a few smiles, and more than one wink. It never looked the same, and in its changes I saw the soul of that anemone, and perhaps the sea itself.

Jeffrey L. Rotman

PURPLE TIP SEA ANEMONE

Beaded Anemone

TEALIA ANEMONE

Most reef-fish avoid these beautiful flower-like animals, and with good reason...These are dangerous blooms which threaten their visitors with a fatal embrace, and yet they have their friends. Almost every anemone was inhabited by some small fish.

Irenäus Eibl-Eibesfeldt

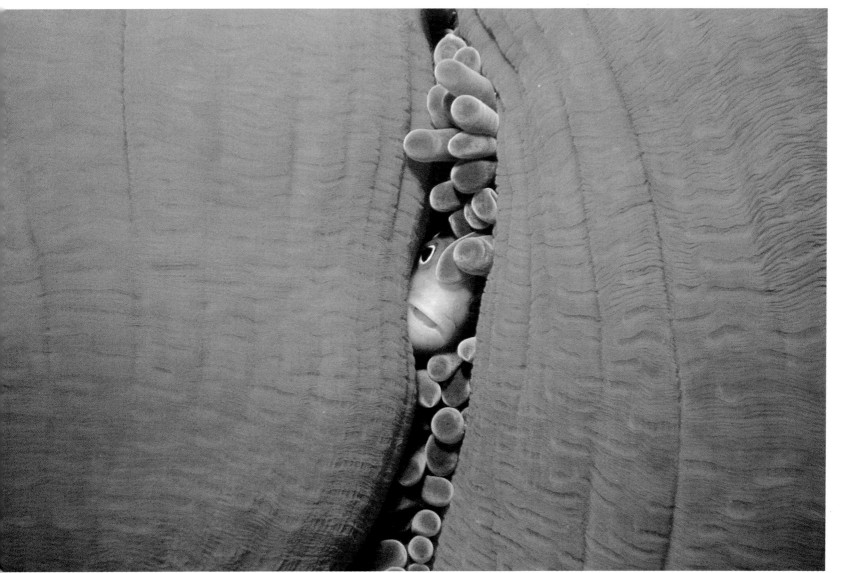
Anemone Fish In Anemone

Orange Coral

It is an old dream: to travel on the back of a benevolent sea beast down to some secret underwater garden.

Stephen Harrigan

KELPFISH IN CORAL

SOCIAL FEATHER DUSTER

Mangroves Under Water

*M*angroves: there is something primeval about them, their geometric tangle of aerial roots, their rich organic smells, their sounds of insects and bubbling muds. They are the base of the food chain for pink shrimp, shelter to lobsters, and hiding place to crocodiles.

Jack Rudloe

PLANTS ENCRUSTED WITH ZEBRA MUSSELS

MALE AND FEMALE COWRIES

The average shell collector may not be fully aware of the valiant battle mollusks have fought and the billions of them that have fallen casualty over millions of years. Through evolution, mollusks and their predators have achieved a dynamic balance. When humans enter the battle on the predators' side, they can rapidly tilt the balance against the mollusks. Humans should not become an enemy in the continuing natural arms race among sea creatures.

Edwin S. Iverson and Darryl E. Jory

... *Y*OU ARE FOUND

UPON THIS SOFT AND SANDY MOUND.
COOLED BY THE SPRAY, ALL SAFE AND SOUND.
AND NOT ONE POINT IN ALL YOUR FIVE
IS EVEN NICKED: YOU SPRAWL ALIVE,
NOT EVEN DENTED BY YOUR DIVE.
BRAVE STAR, I HOPE THAT YOU WILL LIE
LAZILY HERE AND NEVER TRY
TO JUMP BACK UP INTO THE SKY.

WINNIFRED WELLES

OCHRE SEA STARS

Sea Stars

Flamingo Tongue Snail

Pelagic Red Crab

Arrow Crab

I want to realise brotherhood of identity not merely with the beings called human, but I want to realise identity with all life, even with such things as crawl upon the earth.

Mohandes Gandhi

SPINY LOBSTER

DECORATOR CRAB

THE FLOOR OF THE OCEAN IS, IN A SENSE, THE STRANGEST "PLANET" BEING EXPLORED TODAY.

RUTHERFORD PLATT

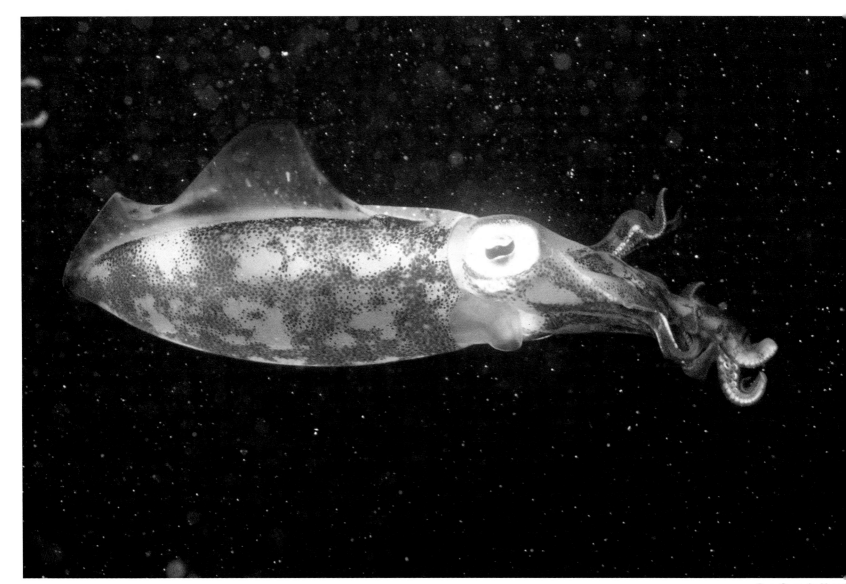
Reef Squid At Night

PACIFIC OCTOPUS

COON STRIPE SHRIMP

We do not associate the idea of antiquity with the ocean, nor wonder how it looked a thousand years ago, as we do of the land, for it was equally wild and unfathomable always.

Henry David Thoreau

AURELIA JELLYFISH

The sea never changes and its works, for all the talk of men, are wrapped in mystery.

Joseph Conrad

Polyorchis Jellyfish

SEA HORSE

CREATURES OF MYTHOLOGY BROUGHT TO LIFE A FISH THAT SHARES THE MARVELOUS CHARACTERISTICS OF A MONKEY AND HORSE, KANGAROO AND CHAMELEON, AND WHICH PROVIDES AN UNUSUAL EXAMPLE OF PATERNAL NURTURE IN THE ANIMAL KINGDOM. THE SEAHORSE WILL ALWAYS BE REGARDED BY NATURALISTS WITH SPECIAL AFFECTION AND A CERTAIN WONDER.

WILLIAM ARRIGONI

Green Moray Eel and Schooling Porkfish

Conger Eel And Coral Shrimp

*I don't mind eels
Except as meals.
And the way they feels.*

Ogden Nash

MANDARIN GOBY

YELLOWHEAD WRASSE

Fish swam by. . .and it seemed again as if they were floating through air with the rocks and seaweed and sea moss like the terrain of some jungle land.

Piers Anthony

GLASSY SWEEPERS

Stoplight Parrotfish

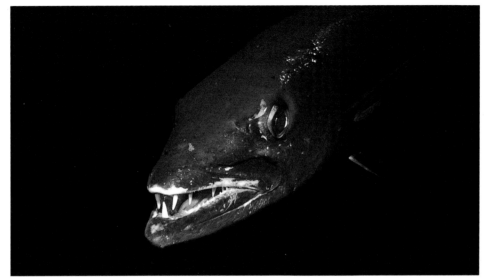

BARRACUDA

By viewing nature, nature's handmaid art
makes mighty things from small beginnings grow:
thus fishes first to shipping did impart
their tail the rudder, and their head the prow.

JOHN DRYDEN

LIONFISH

TRUMPET FISH

PORCUPINE FISH

No water is still, on top.
Without wind, even, it is full
Of a chill, superficial agitation.
It is easy to forget,
Or not to know at all

That fish do not move
By means of this rippling
Along the outside of water, or
By anything touching on air.
Where they are, it is still . . .

James Dickey

STRIPED SNAPPERS

ALLIGATOR GAR

*O*ne way to open your eyes to unnoticed beauty is to ask yourself, "What if I had never seen this before? What if I knew I would never see it again?"

Rachel Carson

HAWKFISH

Spangles of sunlight glitter on the silent blue of the sea. Suddenly, from out of the depths flashes a glistening giant, its huge, sleek body catapulting clear of the riven waters. With a thunderous sound, the giant crashes back to the surface and vanishes beneath it. The sea is silent again.

Harold W. McCormick
and Tom Allen
with Captain William E. Young

Giant Manta Ray

The great fish moved silently through the night water, propelled by short sweeps of its crescent tail. The mouth was open just enough to permit a rush of water over the gills. There was little other motion: an occasional correction of the apparently aimless course by the slight raising or lowering of a pectoral fin—as a bird changes directions by dipping one wing and lifting the other. The eyes were sightless in the black, and the other senses transmitted nothing extraordinary to the small, primitive brain. The fish might have been asleep, save for the movement dictated by countless millions of years of instinctive continuity: lacking the flotation bladder common to other fish and the fluttering flaps to push oxygen-bearing water through its gills, it survived only by moving. Once stopped, it would sink to the bottom and die of anoxia.

Peter Benchley

...The prowling shark, that villainous footpad of the seas, would come skulking along, and, at a wary distance, regard us with his evil eye.

Herman Melville

Caribbean Reef Shark

Paraguayan Caiman

Marine Iguana

It is a hideous-looking creature, of a dirty black colour, stupid, and sluggish in its movements.

Charles Darwin

Previous Page – Great (American) Egret

No bird soars too high,
if he soars with his own wings.

William Blake

Hermann's and California Gulls

Magnificent Frigatebird

Cattle Egret

"Goneys and gullies an' all o' the birds o' the sea,
   They ain't no birds, not really," said Billy the Dane.
"Not mollies, nor gullies, nor goneys at all," said he,
   "But simply the sperrits of mariners livin' again."

   John Masefield

GREAT BLUE HERON

GULL AT SUNSET

. . . They rise above the green

grass and lightly sway on their long pink stems,

side by side, like enormous feathery blossoms,

seducing (more seductively than Phryne)

themselves; till, necks curling, they sink their large

pale eyes into the softness of their down,

where apple-red and jet-black lie concealed.

Rainer Maria Rilke

Another day, having placed myself between a penguin and the water, I was much amused by watching its habits. It was a brave bird; and till reaching the sea, it regularly fought and drove me backwards. Nothing less than heavy blows would have stopped him: every inch he gained he firmly kept, standing close before me erect and determined.

Charles Darwin

MACARONI PENGUIN

GALAPAGOS PENGUIN

Adélie Penguins, Poulet Island, Antarctic

We walked through many colonies of Adélie, Gentoo, and Chinstrap Penguins—some numbering hundreds of thousands of individuals, resplendent in their black and white "formal dress". . . The cacophony in the colonies was incredible as returning penguins went through complex and noisy greeting ceremonies with their mates. . . The comings and goings of the penguins had a certain comic character. The upright shuffling waddle is often interrupted by joyous tobogganing on the belly down snowy slopes. When one couple's attention is distracted by the reunion ceremony, a neighboring penguin may sneak in and steal pebbles for its own nest.

Paul and Anne Ehrlich

Gentoo Penguins

Chinstrap Penguins

Adélie Penguins, Antarctic Peninsul

RIVER OTTER

Otters are extremely bad at doing nothing...There is, I am convinced, something positively provoking to an otter about order and tidiness in any form, and the greater the state of confusion that they can create about them the more contented they feel.

Gavin Maxwell

ALASKAN SEA OTTER

Young River Otters

AUSTRALIAN SEA LION

Ten meters below the chilly surface. . .I failed to see the young sea lions swimming behind me. . .I turned and found myself looking directly into two large, round, inquisitive eyes, puppylike face. The expression of curiosity immediately erupted into a mass of bubbles as the sea lion snapped its jaws and took off, leaving behind a trail of shimmering silver spheres.

Steven Rosenberg

GALAPAGOS SEA LION

WALRUS AND CALF

"The time has come," the Walrus said,
"To talk of many things;
Of shoes—and ships—and sealing wax—
Of cabbages—and kings—
And why the sea is boiling hot—
And whether pigs have wings."

Lewis Carroll

Betwixt the quarters flows a golden sea;
But foaming surges there in silver play;
The dancing dolphins with their tails divide
the glittering waves, and cut the precious tide.

VIRGIL

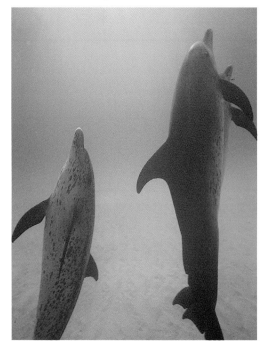

SPOTTED DOLPHINS

These exuberant antics do not seem to be performed to remove parasites or escape predators as the dolphin will leap and spin under any conditions. One dolphin twisting and spinning sets the other off. We could only surmise that they are expressing a pure:"Joie de vivre."

Martha Holmes

Spotted Dolphins

Commerson's Dolphins

It is an important and popular fact that things are not always what they seem. For instance, on the planet Earth, man had always assumed that he was more intelligent than dolphins because he had achieved so much—the wheel, New York, wars and so on—whilst all the dolphins had ever done was muck about in the water having a good time. But conversely, the dolphin had always believed that they were far more intelligent than man—for precisely the same reason.

Douglas Adams

WEST INDIAN MANATEES

YESTERDAY, WHEN I WAS GOING TO THE RIO DEL ORO, I SAW THREE SIRENS COME VERY HIGH OUT OF THE SEA. THEY ARE NOT AS BEAUTIFUL AS THEY ARE PAINTED BECAUSE IN WAYS THEY HAVE A FACE LIKE MAN.

CHRISTOPHER COLUMBUS

False Killer Whale

LEVIATHAN...

UPON EARTH THERE IS NOT HIS LIKE,

WHO IS MADE WITHOUT FEAR.

WILL HE SPEAK SOFT WORDS UNTO THEE?

JOB 41

BELUGA WHALE

Yet calm, enticing calm, oh, whale! thou glidest on, to all who for the first time eye thee...
                                    Herman Melville

BELUGA WHALE WITH CALF

Sperm Whale with Remoras

...And from its pre-historic, main-frame mind,

From its head, twenty-five feet in length,

The Blue Whale,

The largest creature ever to have lived upon this planet,

Now and again utters an ominous whistle

Of a hundred and eighty decibels...

Louder than Concorde.

Heathcote Williams

A mind can only be as healthy as the environment that surrounds it and the environment can only be as healthy as the minds that behold it.

Carlos Eyles

Sunrise, Santa Cruz, CA

To waste, to destroy, our natural resources, to skin and exhaust the land instead of using it so as to increase its usefulness, will result in undermining in the days of our children the very prosperity which we ought by right to hand down to them amplified and developed.

Theodore Roosevelt

The Authors

Adams, Douglas — The Hitchhiker's Guide to the Galaxy by Douglas Adams. © 1979 by Douglas Adams. Published by Harmony.

Anthony, Piers — Robot Adept by Piers Anthony. © 1988. Published by Putnam. Reprinted by permission of The Putnam Publishing Group.

Arrigoni, William — "Seahorses" by William Arrigoni translated by Linda Nemerow Ulman. Sea Frontiers, Nov-Dec 1989.

Benchley, Peter — From the novel Jaws by Peter Benchley. © 1974 by Peter Benchley. Permission granted by International Creative Management, Inc..

The Bible — Job 41.

Blake, William — "Proverbs of Hell" from The Marriage of Heaven and Hell (1790-1793) by William Blake 1757-1827.

Browning, Robert — "Paracelsus" (1835) by Robert Browning 1812-1889.

Carroll, Lewis — Alice Through the Looking Glass, and What Alice Found There (1871) by Lewis Carroll 1832-1898.

Carson, Rachel — The Sense of Wonder by Rachel Carson. © 1956 Rachel Carson. Reprinted by permission of Harper Collins, Inc..

Columbus, Christopher — Journals 1492-1493 by Christopher Columbus 1451-1506.

Conrad, Joseph — Typhoon (1902) by Joseph Conrad 1857-1924.

Darwin, Charles — The Voyage of the Beagle (1845) by Charles Darwin 1809-1882. (pg. 55 refers to a marine iguana).

Dickey, James — Excerpt from "The Movement of Fish" by James Dickey. Reprinted from Drowning With Others. © 1962 by James Dickey, Wesleyan University Press. By permission of University Press of New England.

Dryden, John — "Annus Mirabilis" (1666) by John Dryden 1631-1700.

Durrell, Gerald — Golden Bats and Pink Pigeons by Gerald Durrell. Reproduced with permission of Curtis Brown Ltd. on behalf of Gerald Durrell. © 1977 Gerald Durrell.

Ehrlich, Paul and Anne — Extinction by Paul and Anne Ehrlich. © 1981 Random House. Reprinted by permission of Random House.

Eibl-Eibesfeldt, Irenäus — Land of a Thousand Atolls by Irenäus Eibl-Eibesfeldt. ©1965. Translated from the German by Bwynne Vevers translation © 1965 by MacGibbon and Kee Ltd. World Publishing, Cleveland.

Emerson, Ralph Waldo — Ralph Waldo Emerson 1803-1882.

Eyles, Carlos — Sea Shadows by Carlos Eyles. © 1992 by Carlos Eyles. Reprinted by permission of Watersport Publishers, Inc., P.O. Box 83727, San Diego, CA 92138, 1-800-776-3483.

Gandhi, Mohandes — Mohandes Gandhi 1868-1948.

Harrigan, Stephen — Quotes from Water and Light by Stephen Harrigan. © 1992 by Stephen Harrigan. Reprinted by permission of Houghton Mifflin Company. All rights reserved.

Holmes, Martha — Sea Trek by Martha Holmes, © 1991. Extract reproduced with the permission of BBC Enterprises Ltd.

Iverson, Edwin S. and Jory, Darryl E. — "Arms Race on the Grass Flats" by Edwin S. Iverson and Darryl E. Jory. Sea Frontiers, Sep-Oct 1989.

Masefield, John — From Sea Change (1902) by John Masefield 1878-1967.

Maxwell, Gavin — Ring of Bright Water by Gavin Maxwell. © 1960 by Gavin Maxwell. Reprinted by permission of the publisher, E.P. Dutton & Co.

McCormick, Harold and Allen, Tom with Captain William E. Young — From Shadows in the Sea: The Sharks, Skates and Rays ©1963 by Stein & Day. Reprinted by permission of Scarborough House.

Melville, Herman — Pg. 52 Typee (1846) and pg. 86 Moby Dick (1851) by Herman Melville 1819-1891.

Nash, Ogden — "The Eel" from Verses From 1929 On by Ogden Nash. © 1942 by Ogden Nash. First appeared in The New Yorker. By permission of Little, Brown and Company.

Platt, Rutherford — Water the Wonder of Life by Rutherford Platt. © 1971 by Rutherford Platt. Reprinted by permission of the publisher, Prentice Hall/A Division of Simon & Schuster, Englewood Cliffs, N.J..

Plutarch — Placita Philosophorum by Plutarch c.46AD-c.120AD.

Rilke, Rainer Maria — "Flamingoes" by Rainer Maria Rilke 1875-1926. From The Selected Poetry of Rainer Maria Rilke. Translation © 1982 by Stephen Mitchell. Reprinted by permission from Random House, Inc, NY

Roosevelt, Theodore — Excerpt from Message to Congress December 3, 1907 by Theodore Roosevelt 1858-1919.

Rosenberg, Steven — "The Sea Lions of Monterey" by Steven Rosenberg. Sea Frontiers, Mar-Apr 1989.

Rotman, Jeffrey L. — From the book Colors of the Deep by Jeffrey L. Rotman. Published by Thomasson-Grant, Charlottesville, VA. World copyright 1990 by Edizioni White Star, Vercelli, Italy.

Rudloe, Jack — The Wilderness Coast: Adventures of a Gulf Coast Naturalist by Jack Rudloe. ©1988 E.P. Dutton. Reprinted by permission of the publisher E.P. Dutton.

Scarborough-Bull, PhD, Ann — Creatures, Corals, and Colors in America's Seas by Ann Scarborough-Bull, PhD. © 1990 Audubon Park Press.

Sefton, Nancy — "The Secret Lives of Sponges" by Nancy Sefton. Sea Frontiers, May-Jun 1989.

Thoreau, Henry David — Cape Cod by Henry David Thoreau 1817-1862 (Published posthumously c 1892).

Virgil — The Aenid (29BC-19BC) by Virgil 70BC-19BC.

Welles, Winnifred — "Starfish" by Winnifred Welles. d.1939.

Whitman, Walt — Leaves of Grass (1855) by Walt Whitman 1819-1892.

Williams, Heathcote — Whale Nation by Heathcote Williams. © 1988 by Heathcote Williams. Published by Harmony.

The Photographers

Frank Balthis
Natural history and travel photographer Frank Balthis has photographed our planet for over 20 years. A former national park ranger, Frank's work has been published by the National Geographic Society, National Wildlife Federation, Sierra Club and others. Frank is especially well-known for his coverage of marine mammals and the California coast.
Photos: Pages 7, 15, 16 top, 26, 58, 61 right, 91.

Mark Conlin
Mark Conlin holds a degree in marine biology and has worked on two award-winning underwater wildlife documentaries for Howard Hall Productions. Mark's work has appeared in International Wildlife, Smithsonian, and Natural History, among others.
Photos: Pages 17, 18, 21 left, 28, 29 middle, 31 top, 32, 33, 34, 35, 37, 49, 77, 78, 79, 89.

Michele and Howard Hall
Natural history wildlife film producer, cinematographer, still photographer and writer, Howard Hall has won many prestigious awards including five Emmys. He is the author of HOWARD HALL'S GUIDE TO SUCCESSFUL UNDERWATER PHOTOGRAPHY. Former registered pediatric nurse Michele Hall has been taking underwater still photographs since 1977. Howard and Michele co-produced a 1994 National Geographic special, "JEWELS IN THE CARIBBEAN."
Photos: Pages 10 left, 11, 19, 33 right, 43, 45, 53, 71, 87.

Robert Rattner
Bob's documentary journalism and travel, nature and underwater photography have appeared in Audubon, International and National Wildlife, National Geographic, The New York Times Magazine, and Smithsonian. In 1991 Bob became the first recipient of the Marty Forscher Grant, awarded for excellence in socially important photography.
Photos: Pages 9, 13, 23, 30, 33 right, 54, 55 bottom, 56-57, 59, 60, 61 left, 64 right, 70, 82, 83, 92.

David and Carol Sailors
A graduate of Indiana University, David has exhibited throughout the U.S. Among his awards are the Award of Distinction, Society for Technical Communications, first place PHOTO National Juried Exhibition, and the Barbara Hershey Alternative Processes Award. Carol, a lyric and nonfiction writer, has photographed coral reefs in the Caribbean, Florida Keys and Papua New Guinea.
Photos: Pages 21 right, 22, 42, 44 middle and right, 46.

Charles Seaborn
Underwater photographer Charles Seaborn was educated at the University of Puget Sound, Hopkins Marine Station of Stanford University, Woods Hole Marine Biology Lab, and the Brooks Institute of Photography. He is involved with a variety of projects that foster an awareness of the marine environment and is the author of UNDERWATER PHOTOGRAPHY.
Photos: Pages 12, 31 bottom.

Kevin and Cat Sweeney
Residents of the Big Island of Hawaii, the Sweeneys specialize in nature photography of the underwater environment. Their award-winning images range from intricate macro subjects to humpback whales and dolphins. They have been published worldwide in numerous books, magazines, CDs, calendars, and posters.
Photos: Pages Cover, Endpapers, 4, 24 left, 39, 41, 47 bottom, 74-75.

Robert Trifone
Underwater photographer Rob Trifone holds both a bachelor's and a master's degree in biology. He teaches biology at Brien McMahon High School in Norwalk, Connecticut, and teaches a course in tropical marine biology in the Caribbean.
Photos: Pages 10 right, 14, 16 bottom, 20, 25, 29 left and right, 40 right.

The Photographers of Sea World
Known for their superb technique as well as for their dedication to the marine environment, Sea World's photographers are world renowned. Their award-winning work has appeared in A DAY IN THE LIFE OF CALIFORNIA, People, Sports Illustrated, and on the AP wire service.
Photos: Pages 27, 33 left, 36, 37, 38, 40 left, 44 left, 47 top, 50, 52, 62, 64 left, 65, 66, 67, 68, 69, 72, 80, 84, 85, 86.

Ken Bohn — California
Bob Couey — California
Bob French — Texas
Chris Gotshall — Florida
Steve Szerdy — Florida

Postlude
A personal observation

I live on a pond.
The pond is connected to the sea, in the same way that I am connected to you, and we are all connected to an ancestral past that began with the first drop of water that blessed a dry and inhospitable planet.
When it comes to the sea, there are no six degrees of separation. We are bound to it, and as we hurtle towards an increasingly fragmented world, the sea is the great unifier.
When I look out to the ocean, it is with the same wonder that the child in a coastal village gazes to the horizon.
Then why are we destroying it? We call ourselves more civilized than the "primitives" who worshipped the sea. Yet we literally trash it. Ocean dumping continues, and motivated by raw greed renegades pump raw sewage by the thousands of tons.
More advanced technology can mean more advanced means of destruction.
In the end, how do you judge a civilization? By the way it treats children, and women, and animals — and by the way it treats the sea.
We need a Secretary of the Sea, a new cabinet-level position.
We need to take the magnificent photographs in this book and hang copies in the conference room of every corporation in America as a vivid reminder that the oceans must be preserved.

We need to bring up a generation of children that are taught to see below the surface, and to love and humor what's there.
I read recently that astronomers, for the first time, have found signs of water in a distant galaxy named Markarian 1. It is the farthest point in the universe where water has been detected. The scientists believe that the presence of water suggests that life might exist 200 million light-years away.
If that's true, I hope that they're doing a better job of cherishing their oceans and seas than we are.
Will we learn? I believe we will. My work is about looking into the future to understand change and to prepare us for it. I see a great movement towards taking responsibility for the fate of our world. A growing connection that recognizes a common obligation to stop the madness.

I live on a pond.
The pond is connected to the sea, in the same way that you are connected to me.

Faith Popcorn